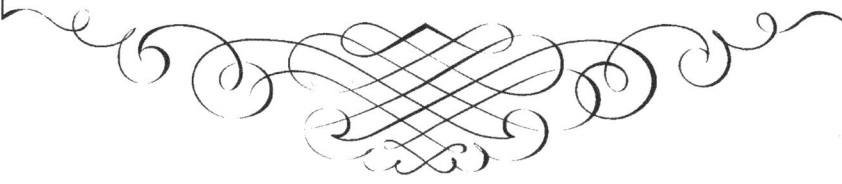

ISBN 978-0-428-00943-4
PIBN 11167323

This book is a reproduction of an important historical work. Forgotten Books uses
state-of-the-art technology to digitally reconstruct the work, preserving the original format
whilst repairing imperfections present in the aged copy. In rare cases, an imperfection in
the original, such as a blemish or missing page, may be replicated in our edition. We do,
however, repair the vast majority of imperfections successfully; any imperfections that
remain are intentionally left to preserve the state of such historical works.

English
Français
Deutsche
Italiano
Español
Português

www.forgottenbooks.com

Mythology Photography **Fiction**
Fishing Christianity **Art** Cooking
Essays Buddhism Freemasonry
Medicine **Biology** Music **Ancient**
Egypt Evolution Carpentry Physics
Dance Geology **Mathematics** Fitness
Shakespeare **Folklore** Yoga Marketing
Confidence Immortality Biographies
Poetry **Psychology** Witchcraft
Electronics Chemistry History **Law**
Accounting **Philosophy** Anthropology
Alchemy Drama Quantum Mechanics
Atheism Sexual Health **Ancient History**
Entrepreneurship Languages Sport
Paleontology Needlework Islam
Metaphysics Investment Archaeology
Parenting Statistics Criminology
Motivational

Historic, Archive Document

Do not assume content reflects current
scientific knowledge, policies, or practices.

Animal and
Plant Health
Inspection
Service

Plant
Protection and
Quarantine

APHIS 81-47
January 1986

Introduction i
Acknowledgments ii
Key to Families 1
Hesperiidae 10
Pieridae 10
Lycaenidae 12
Geometridae. 12
Noctuidae 13
Carposinidae 17
Pyralidae 18
Tortricidae 41
Cochylidae 45
Blastobasidae. 45
Cossidae 46
Argyresthiidae 47
Cosmopterigidae 47
Oecophoridae 47
Gelechiidae 48
Acrolepiidae 52
Plutellidae 52
Tineidae 53
Selected References 57
Index 61

The following keys are intended to aid in recognizing the
lepidopterous larvae most frequently intercepted at
United States ports of entry. This paper is essentially an
expansion of Capps' keys used by quarantine inspectors since
1939. It includes the 50 species in Capps (1963) and 40
additional species, mostly from the Old World. The characters
have been reworked to accommodate the added species, and
the Heinrich system of setal nomenclature has been replaced by
the more generally used Hinton (1946) system.

These keys are based on mature larvae and may not work for some
of the earlier instars. The characters used to separate the
families are not diagnostic for the families but are intended
only to separate the included species. The host and distribution should be considered in making a determination.

Many of the drawings are diagrammatic, particularly the setal
maps, and are intended only to illustrate characters referred
to in the text. In examining setal maps showing lateral views,
the head of the larva is to the left; in dorsal and ventral
views, the head is toward the top of the figure. In figures
showing crochets, the meson is to the left and the head is
toward the top of the figure.

3. L. Godfrey, Illinois Natural History Survey;
iges and J. M. Kingsolver, Systematic Entomology
ry, U.S. Department of Agriculture; and members of the
nd Plant Health Inspection Service, Plant Protection
antine Training Center, USDA, for suggestions and
ons.

ebted to Molly K. Ryan for inking the line drawings.

1.   Body with numerous short secondary setae, primary setae not evident (fig. 1) 2

1'   Body without numerous short secondary setae, primary setae evident (fig. 2) 4

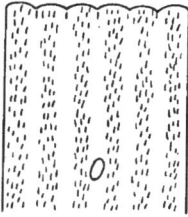

1

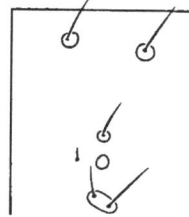

2

2(1).   Head much larger than prothorax (fig. 3); prothorax shorter than succeeding segments (fig. 3); crochets biordinal or triordinal in a laterally elongated circle (fig. 4); body widest at middle . . . . HESPERIIDAE p. 10

2'.   Head equal to the diameter of body or much smaller; prothorax not shorter than succeeding segments; crochets in a mesoseries (fig. 5); body depressed or cylindrical 3

3

4

5

3(2').   Body cylindrical, not depressed; abdominal segments divided into 6 or fewer annulets (fig. 6); head about size of prothorax (fig. 7); crochets in continuous mesoseries (extra short series of crochets may be present) (fig. 8) . . . . . .
. . . . PIERIDAE p. 10

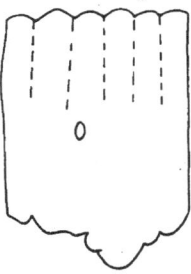 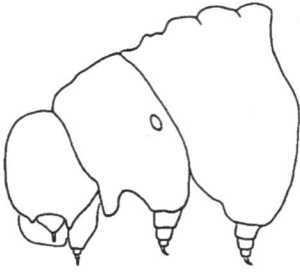 

     6   7   8

3'.   Body depressed, spindle shaped; abdominal segments not divided into annulets; head about one-third size of prothorax, retractile (fig. 9); crochets in mesoseries interrupted at center by spatulate lobe (fig. 10) . . . .
. . . . LYCAENIDAE p. 12

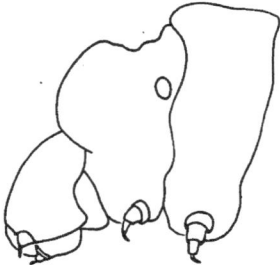 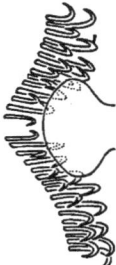

        9   10

4(1').   Prolegs present only on abdominal segments 6 and 10
. . . . GEOMETRIDAE p. 12

4'.   Prolegs present on more abdominal segments 5

5(4').   Two setae in the prespiracular group of the prothorax (fig. 11) 6

5'.   Three setae in the prespiracular group of the prothorax (fig. 12) 8

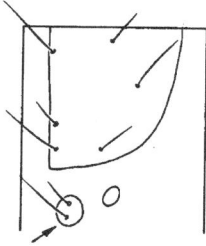 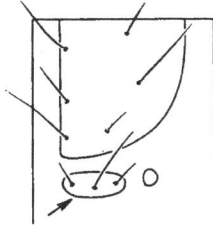

           11   12

6(5).   Abdominal segments 1 to 8 with seta L1 behind and seta L2 below the spiracle (fig. 13); subdorsal setae below the prothoracic shield (fig. 14); crochets in a mesoseries (fig. 15) . . . . NOCTUIDAE p. 13

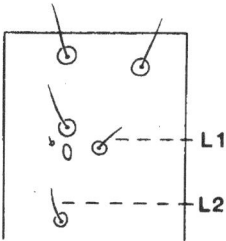 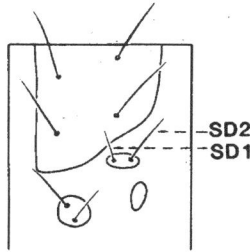 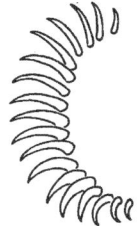

           13   14   15

6'.   Abdominal segments 1 to 8 with setae L1 and L2 close together below the spiracle (fig. 16); subdorsal setae on the prothoracic shield (fig. 11); crochets in a complete circle (fig. 17) or a penellipse (fig. 18) 7

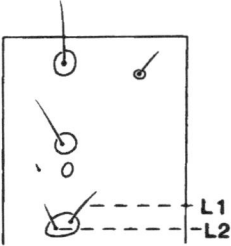

16

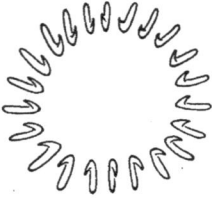

17

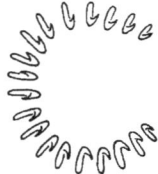

18

7(6').   Four subventral setae present on abdominal segments 3 to 6 (fig. 19); spiracles on abdominal segment 8 well above level of those on preceding segments (fig. 20) . . . .
. . . . CARPOSINIDAE p. 17

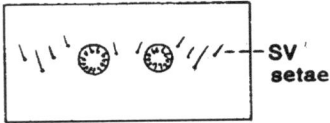

19

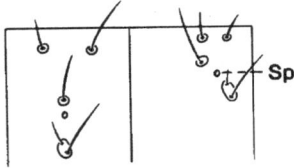

A7   A8

20

7'.   Three subventral setae present on abdominal segments 3 to 6 (fig. 21); spiracles on abdominal segment 8 on a level with those of preceding segments (fig. 22) . . . .
. . . . PYRALIDAE p. 18

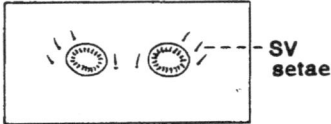

21

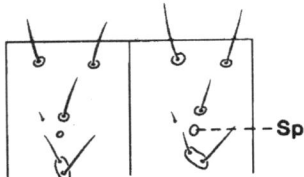

A7   A8

22

8(5'). Setae L1 and L2 of abdominal segments 3 to 6 close together below the spiracle, often on the same pinaculum (fig. 23) 9

8'. Setae L1 and L2 of abdominal segments 3 to 6 widely separated below the spiracle or below and behind the spiracle (fig. 24) 16

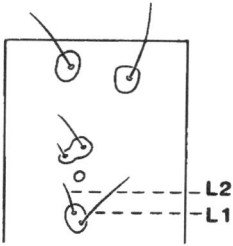

23

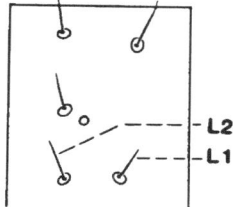

24

9(8). Paired setae D2 of abdominal segment 9 on a single pinaculum (fig. 25) 10

9'. Paired setae D2 of abdominal segment 9 not on a single pinaculum (fig. 26) 11

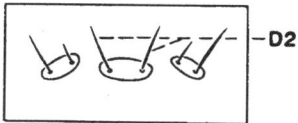

25

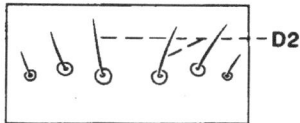

26

10(9).   Three lateral setae present on abdominal segment 9
(fig. 27) . . . . TORTRICIDAE p. 41

10'.   Two lateral setae present on abdominal segment 9
(fig. 28) . . . . COCHYLIDAE p. 45

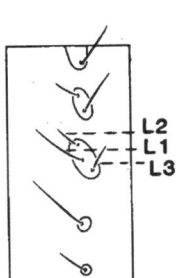

27

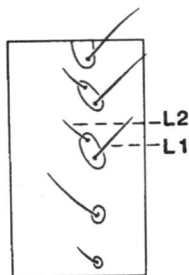

28

11(9').   Submentum with a large oval pit (fig. 29); seta SD1
of abdominal segment 8 above and slightly behind the spiracle
(fig. 30); abdominal segments 1 to 7 often with sclerotized
ring around seta SD1 . . . . BLASTOBASIDAE p. 46

11'.   Submentum without an oval pit; or, if pit is present,
seta SD1 of abdominal segment 8 above and in front of spiracle
(fig. 31) 12

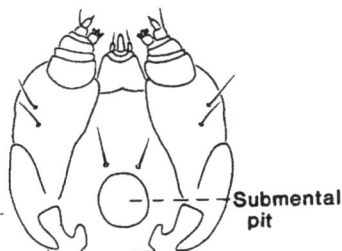

29

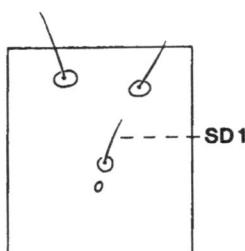

30

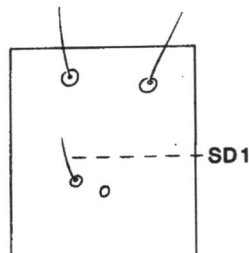

31

(11'). Proleg-bearing segments (3 to 6) of abdomen with an
tra unnamed seta behind the spiracle (fig. 32) . . . .
. . . . COSSIDAE p. 46

'.   Proleg-bearing segments of abdomen without an extra
ta behind the spiracle (fig. 33) 13

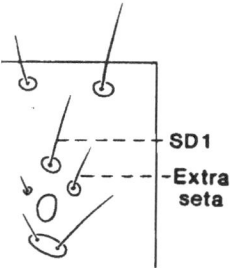

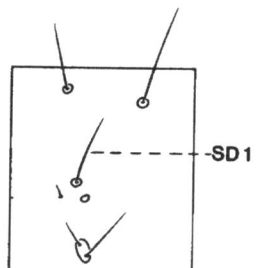

       32   33

(12'). Prothoracic shield with seta SD2 slightly behind line
seta SD1 and XD2, D2 behind seta D1 and much closer to
lline than to lateral margin (fig. 34); abdominal segment 9
th setae D1, D2, and SD1 on common pinaculum (fig. 35) . . .
. . . . ARGYRESTHIIDAE p. 47

'.   Prothoracic shield with seta SD2 well behind line of
tae SD1 and XD2, seta D2 below seta D1 and approximately
lway between midline and lateral margin (fig. 36); abdominal
gment 9 with setae D1, D2, and SD1 on 2 or 3 separate
iacula (figs. 37, 38, 39) 14

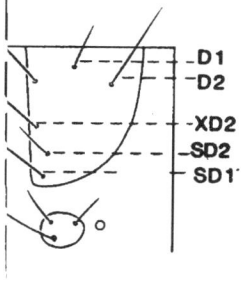

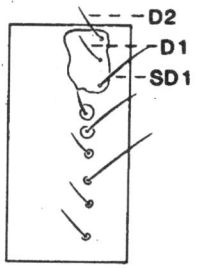

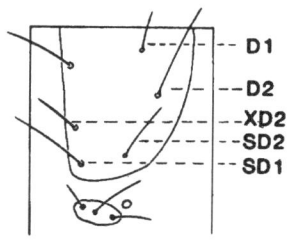

       34   36

                        35

14(13'). Abdominal segment 9 with setae D1 and SD1 on common pinaculum, setae L1 and L2 on separate common pinaculum, seta L3 on its own pinaculum (fig. 37) . . . . COSMOPTERIGIDAE p. 47

14'. Abdominal segment 9 with setae D1 and SD1 not closely associated, not on same pinaculum (figs. 38, 39) 15

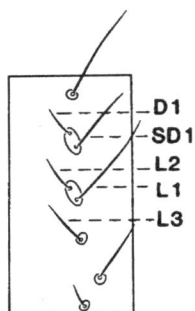

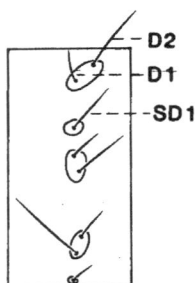

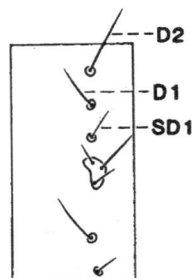

37   38   39

15(14'). Abdominal segment 9 with seta D1 closely associated with and forward of seta D2 (fig. 38); abdominal segment 1 may have 2 or 3 setae in subventral group . . . OECOPHORIDAE p. 48

15'. Abdominal segment 9 with seta D1 equidistant from setae D2 and SD1, usually the three setae in line (fig. 39); abdominal segment 1 always with 2 setae in subventral group (fig. 40) . . . . GELECHIIDAE p. 49

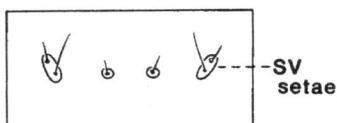

40

16(8'). Pinaculum of seta SD1 enclosing spiracle on a
segments 1 to 8 (fig. 41); crochets of abdominal proleg
uniordinal circle enclosing a short longitudinal series
crochets (fig. 42) . . . . ACROLEPIII

16'. Pinaculum of seta SD1 on abdominal segments 1
not enclosing spiracle; crochets in circle or ellipse v
enclosed series (figs. 43, 44, 45) . . . .

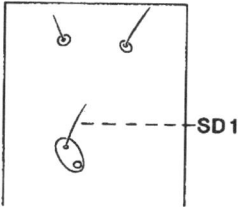

41

42

17(16'). Crochets of abdominal prolegs in a biserial
(fig. 43); seta L3 missing on abdominal segment 9 (fig
. . . . PLUTELLID

17'. Crochets of abdominal prolegs in a uniserial
or ellipse, rows of very small spinules may be present
prolegs anterior or posterior to the crochets (figs. 4
seta L3 usually present on abdominal segment 9 (missin
Tineola) (figs. 218, 220) . . . . TINEID

43

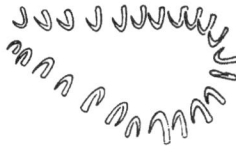

44

4

Distribution: Worldwide
Hosts: citrus, canna, palm spp., and many other plants

1.   Body with soft vestiture of very fine white setae borne on small to very small chalazae (fig. 46); body green 2

1'.   Body with a few very large, broad-based, seta-bearing chalazae interspersed among numerous small setae (fig. 47); body yellow and brown 3

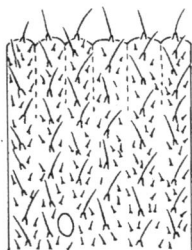

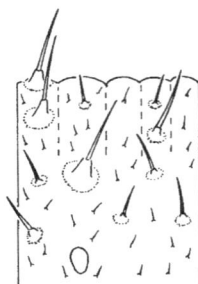

46   47

2(1).   Body with yellow middorsal stripe and a broken stripe through the spiracles (fig. 48) . . . . Pieris rapae (Linnaeus)

   Distribution: Europe and North America
   Hosts: cabbage, cauliflower, and other crucifers

2'.   Body without middorsal stripe, with yellowish area around abdominal spiracles (fig. 49) . . Pieris napi (Linnaeus)

   Distribution: Europe and North America
   Hosts: mustard and turnip

Head black, except for gray front and light patch on
e; body with yellow middorsal and spiracular stripes,
veen with patches of dark color (fig. 50); anal shield
th median yellow stripe . . **Pieris** **brassicae** (Linnaeus)

lbution: Europe, Middle East, North Africa, and Chile
: cabbage, cauliflower, and other crucifers

Head yellow except for black chalazae; body with
ɔrsal, subdorsal and spiracular stripes, area between
ʋo with longitudinal fuscous stripe (fig. 51); anal
ɛllow except for black chalazae . . . .
. . . . **Ascia** **monuste** (Linnaeus)

lbution: Mexico, West Indies, and United States
: cabbage and other crucifers

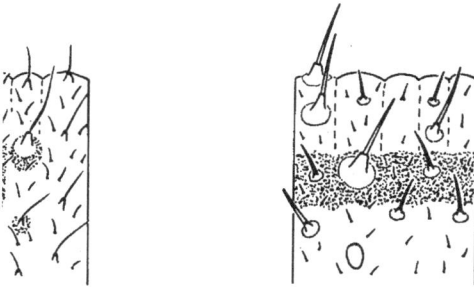

0   51

Keys for Lepidopterous Larvae - 11

Secondary setae long, stout, and of variable length,
₂ on short, cylindrical chalazae (fig. 52) . . . .
. . . . Strymon melinus (Hübner`

Distribution: Mexico and United States
osts: beans and cotton

Secondary setae of moderate, more uniform length,
₂ on short star-shaped chalazae (fig. 53) . . . .
. . . . Lampides boeticus (Linnaeus)

Distribution: Old World and Hawaii
osts: beans, peas, and other legumes

52

53

granulose; setae spatulate (fig. 54) . . . . . Idaea spp.

Distribution: Mexico and Europe
osts: cut flowers, heather, and chamomile

54

Prolegs absent on abdominal segments 3 and 4, present
egments 5 and 6 2

Prolegs present on abdominal segments 3 to 6 . . . . ɔ

.   Vestigial prolegs present on abdominal segments 3 and
etae SV1, SV2, and V1 grouped closely about vestigial
eg (fig. 55); pinacula of setae SV1 and SV2 well separated
bdominal segment 2 (fig. 55) . . . <u>Trichoplusia ni</u> (Hübner)

istribution: North, Central, and South America and
West Indies
osts: general feeder

Vestigial prolegs absent on abdominal segments 3 and
eta V1 separated from setae SV1 and SV2 on all abdominal
ents (fig. 56); pinacula of setae SV1 and SV2 fused on
minal segments 2 to 4 (fig. 56) . . . .
· · · · · · · · · · · · · · · <u>Autographa gamma</u> (Linnaeus)

istribution: Europe, Asia, and North Africa
osts: general feeder

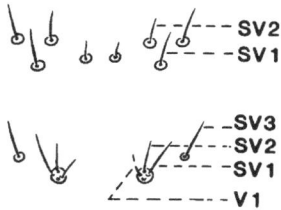

55

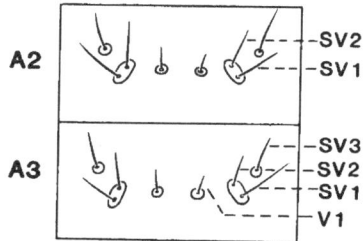

56

3(1').   Integument with short, sharp spines (fig. 57) . . . 4
3'.   Integument smooth or with flat granules (fig. 58)
. . . . 5

57

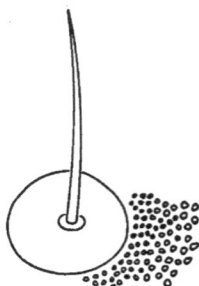

58

4(3).   Chalazae D1 and D2 of abdominal segments with hairlike spinules (fig. 59); mandible with a broad basal process on
the oral surface (fig. 60) . . *Heliothis virescens* (Fabricius)

    Distribution: Mexico, West Indies, and United States
    Hosts: cotton, tobacco, tomatoes, peppers, and many others

4'.   Chalazae D1 and D2 of abdominal segments without
spinules (fig. 61); mandible without basal process on the oral
surface (fig. 62) . . . . **Helicoverpa zea** (Boddie)

    Distribution: North, Central, and South America, and
    West Indies
    Hosts: corn, beans, cotton, tomatoes, and many others

59

60

61

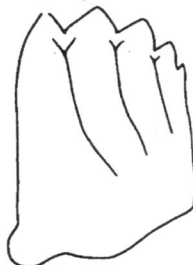

62

5(3'). Meso- and metathorax with dark bar connecting seta SDI with adjacent ventral muscle attachment, no dark bar associated with seta SD2 on these segments (fig. 63) 6

5'. Meso- and metathorax with dark bars connecting both setae SD2 and SDI with their adjacent ventral muscle attachments (fig. 64) 7

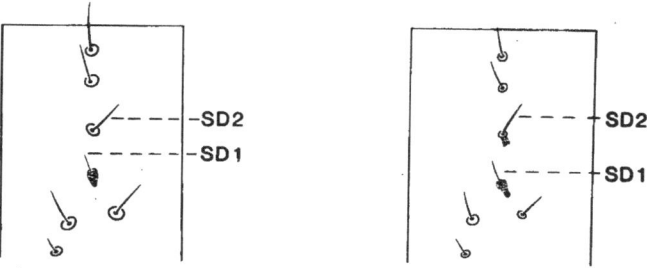

63   64

6(5). Integument with evident flat granules (fig. 65); setal pinacula of body large and brown (fig. 66); abdominal segments patterned as illustrated (fig. 66) . . . .
. . . . Spodoptera frugiperda (J. E. Smith)

   Distribution: North, Central, and South America, West Indies
   Hosts: wide range of crop plants and vegetables

6' Integument smooth; setal pinacula of body minute (fig. 67); abdominal segments patterned as illustrated (fig. 67) . . . . Spodoptera exigua (Hübner)

   Distribution: Old World, Mexico, and United States
   Hosts: wide range of vegetables

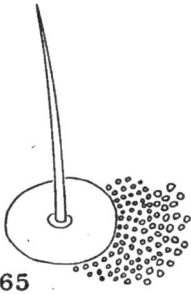
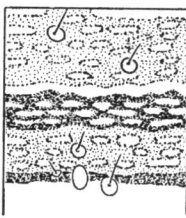
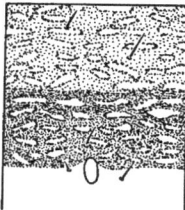

65   66   67

7(5'). Mandible with large molar-bearing basal process on oral surface (fig. 68) . . . . . **Mamestra brassicae** (Linnaeus)

Distribution: Europe and Asia
Hosts: crucifers and other leafy vegetables

7'. Mandible without large basal process (fig. 69) (a small tooth may be present at base of oral surface in **Xestia**) 8

69

68

8(7'). Spinneret with apical spinules (fig. 70); spiracles black; yellow middorsal spots on metathorax and abdominal segments 1 to 4 . . . . **Peridroma saucia** (Hübner)

Distribution: Europe, North, Central, and South America, and West Indies
Hosts: general feeder

8'. Spinneret without spinules; spiracles white or yellowish; no yellow middorsal spots 9

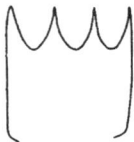

70

').   Combined length of 2nd and 3rd segments of labial
pus one-half or more length of 1st segment (fig. 71);
nneret with 2 denticles on apical margin (fig. 71); head
iculated brown usually with black submedian arcs present
. . . . <u>Xestia</u> spp.

Distribution: Europe and North America
Hosts: general feeder

      Combined length of 2nd and 3rd segments of labial
pus one-third or less length of 1st segment (fig. 72);
nneret without denticles, apical margin lobed (fig. 72);
er instar head reticulated with light brown, no darker
median arcs, head flecked with brown in early instars;
. . . . <u>Copitarsia</u> spp.

Distribution: Mexico, Central and South America
Hosts: cut flowers, potatoes, and other vegetables

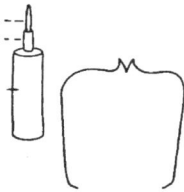

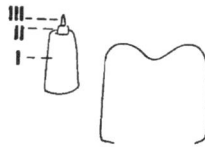

71

72

      family never has more than one lateral seta on abdominal
ent 9 . . . . . <u>Carposina</u> <u>niponensis</u> <u>niponensis</u> Walsingham

istribution: Japan, Korea, and China
oats: apple, pear, plum, and peach

1. Sclerotized ring around seta SD1 on abdominal segment
8 (missing in _Etiella zinckenella_) (fig. 73); 3 setae in the
lateral group on abdominal segment 9 (fig. 74) 2

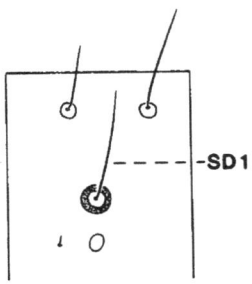

73

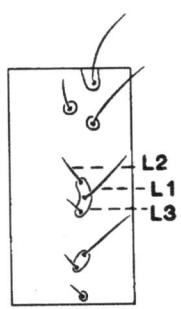

74

1'. No sclerotized ring around seta SD1 on abdominal
segment 8 (fig. 75); one seta in the lateral group on
abdominal segment 9 (fig. 76) . . . . 20 (p. 30)

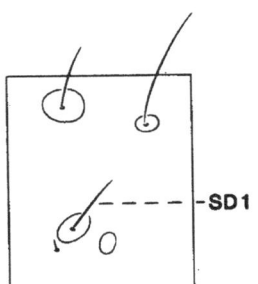

75

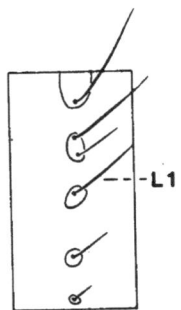

76

2(1).   Sclerotized ring around seta SD1 on mesothorax
(fig. 77) . . . . Phycitinae . . . 3

2'.   No sclerotized ring around seta SD1 on mesothorax
. 15

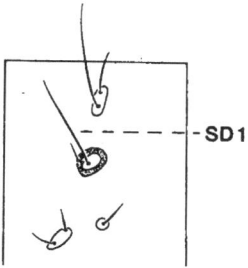

77

3(2).   Prespiracular shield of prothorax extending below and
behind the spiracle, posterior portion weakly pigmented (fig.
78); body pink with whitish discontinuous longitudinal bands on
most segments (fig. 79) . . . Elasmopalpus lignosellus (Zeller)

   Distribution: Mexico, West Indies, and United States
   Hosts: corn, sugarcane, peas, and many others

3'.   Prespiracular shield of prothorax never extending
below and behind spiracle (fig. 80); body white or if pinkish
without white bands 4

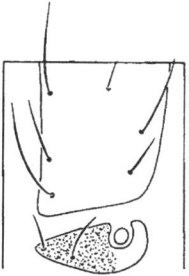
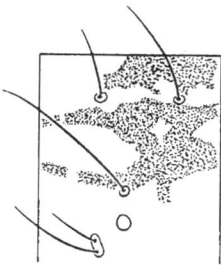
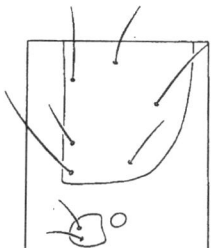

78   79   80

4(3').   Integument granulose under low magnification (30X)
(fig. 81) 5

4'.   Integument not granulose under low magnification
. . . . 7

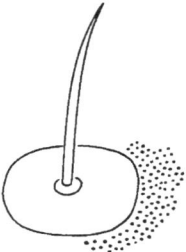

81

5(4).   Prothoracic shield with black areas on lateral
margins and longitudinal black areas on either side midway
between center line and lateral margins (fig. 82) . . . . . .
. . . . Ancylostomia stercorea (Zeller)

    Distribution: West Indies
    Hosts: pigeon peas

5'.   Prothoracic shield not with the above color pattern
. . . . 6

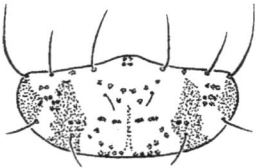

82

6(5').   Pinacula of body setae large and dark (fig. 83); seta
D2 of abdominal segments 1 to 7 below level of seta D1 (fig.
83) . . . . **Hypsipyla grandella** (Zeller)

   Distribution: Central and South America, West Indies, and
     southern Florida
   Hosts: mahogany and Spanish cedar logs

6'.   Pinacula of body setae very small and pale (fig. 84);
seta D2 of abdominal segments 1 to 7 at level of seta DI (fig.
84) . . . . **Moodna bisinuella** Hampson

   Distribution: Mexico
   Host: corn

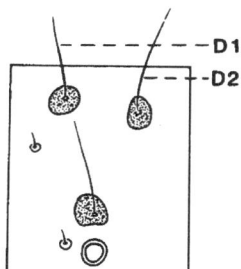 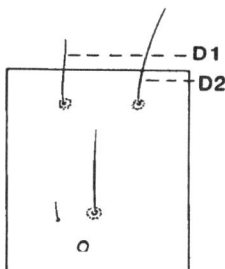

      83   84

7(4').   Prothoracic shield yellow with pattern of dark marks
as illustrated (fig. 85) . . . . . . **Fundella pellucens** Zeller

   Distribution: Mexico and West Indies
   Hosts: beans and peas

7'.   Prothoracic shield yellowish without the pattern as
above 8

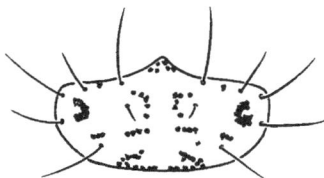

      85

8(7').   Coronal suture absent (fig. 86); abdominal segments
1 to 7 with a crescent-shaped patch above seta SD1 (usually
reduced to a small smudge or missing in _Amyelois transitella_)
(fig. 87) 9

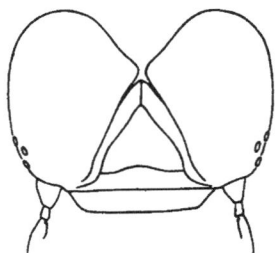

86

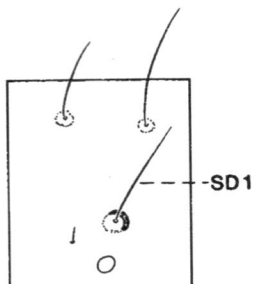

87

8'.   Coronal suture present (fig. 88); abdominal segments
1 to 7 without crescent-shaped patch above seta SD1 (fig. 89)
. . . . 10

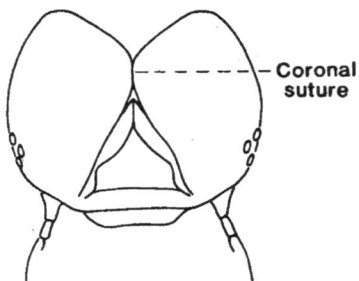

88

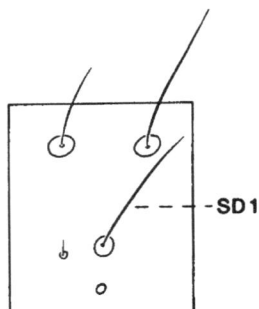

89

9(8).   Anal plate with seta SD1 closer to seta D1 than to
seta L1 (fig. 90); seta SD2 of abdominal segment 8 usually
separated from the spiracle by 2 or more times the diameter of
the spiracle (fig. 91); ring around seta SD1 on abdominal
segment 8 usually complete (fig. 91) . . . .
. . . . Ectomyelois ceratoniae (Zeller)

   Distribution: Mediterranean, Africa, Argentina, West Indies,
   and Florida
   Hosts: legumes, nuts, dates, tamarinds, carobs, and others

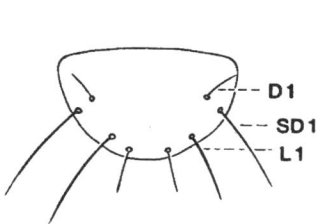

90

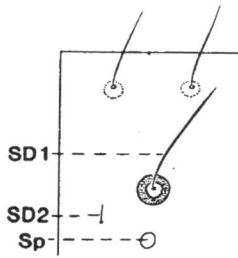

91

9'.   Anal plate with seta SD1 equidistant from setae D1
and L1 (fig. 92); seta SD2 of abdominal segment 8 usually
separated from the spiracle by one to 1.5 times the diameter of
the spiracle (fig. 93); ring around seta SD1 on abdominal
segment 8 incomplete (fig. 93) . . . .
. . . . Amyelois transitella (Walker)

   Distribution: North and South America, and West Indies
   Hosts: oranges, walnuts, and other fruits and pods

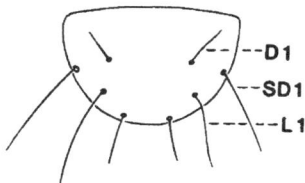

92

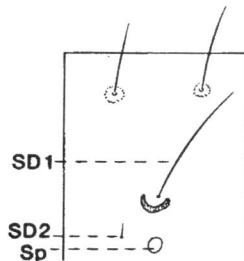

93

10(8').　Abdominal segments 1 to 8 apparently without pinacula (pinacula concolorous with body and not evident) (fig. 94) . . . . **Plodia** **interpunctella** (Hübner)

　　Distribution: Cosmopolitan
　　Hosts: stored grain, vegetable and fruit products

10'.　　Abdominal segments 1 to 8 with small pigmented pinacula (fig. 95) 11

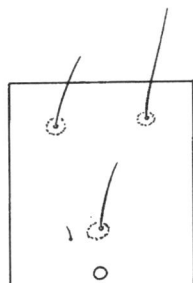 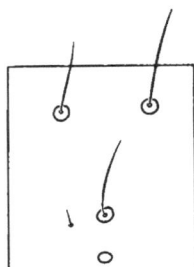

94　　　　　　　　　　95

11(10').　Abdominal segment 8 with seta SD2 separated from spiracle by 2 to 3 times the horizontal diameter of the spiracle (fig. 96) 12

11'.　　Abdominal segment 8 with seta SD2 separated from spiracle by a distance equal to the horizontal diameter of the spiracle (fig. 97) 13

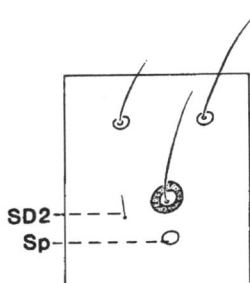 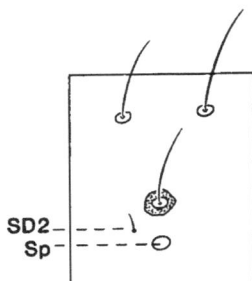

96　　　　　　　　　　97

12(11).   Spiracle of abdominal segment 8 as large as the area
enclosed by the sclerotized ring around seta SDI (fig. 98) . .
· · · · · · · · · · · · · · · · · · · Anagasta kuehniella (Zeller)

   Distribution: Nearly cosmopolitan
   Hosts: grain and other stored and dried vegetable products

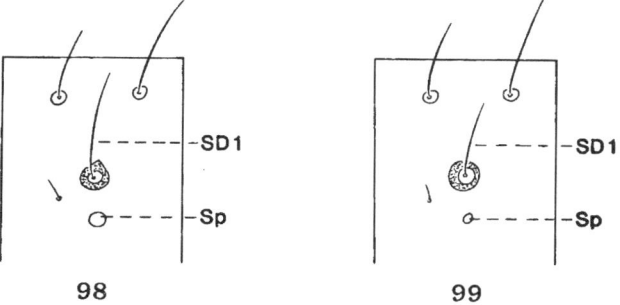

12'.   Spiracle of abdominal segment 8 two-thirds or less
as broad as the area enclosed by the sclerotized ring around
seta SDI (fig. 99) · · · · · · · · · Ephestia elutella (Hübner)

   Distribution: Nearly cosmopolitan
   Hosts: stored and dried vegetable products

98

99

13(11').   Seta D2 of abdominal segments 1 to 8, two to two and
one-half times the length of seta DI (fig. 100) · · · · · · ·
· Cadra cautella (Walker)

   Distribution: Cosmopolitan
   Hosts: stored and dried vegetable products

13'.   Seta D2 of abdominal segments 1 to 8, three to five
times the length of seta D1 (fig. 101) · · · · · · · · · · 14

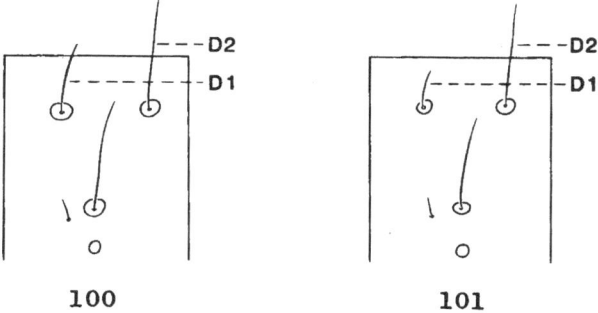

100

101

14(13'). Metathorax with the distance between setae V1 2 times or less than the distance between seta V1 and the coxa (fig. 102) . . . . Cadra figulilella (Gregson)

    Distribution: Nearly cosmopolitan
    Hosts: dried fruits, nuts, seeds, and beans

14'.   Metathorax with the distance between setae V1 3 to 5 times the distance between seta V1 and the coxa (fig. 103) . . . . Cadra calidella (Guenée)

    Distribution: Mediterranean
    Hosts: carobs, dried fruits, and nuts

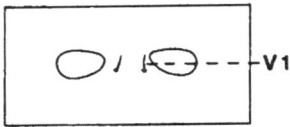

102

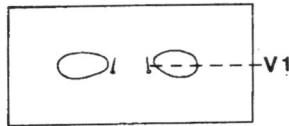

103

15(2'). Prothoracic shield with pattern of dark markings as illustrated (fig. 104) . . . . Phycitinae . . . . Etiella zinckenella (Treitschke)

    Distribution: Nearly cosmopolitan
    Hosts: lima beans, pigeon peas, and other legumes

15'.   Prothoracic shield not patterned as above 16

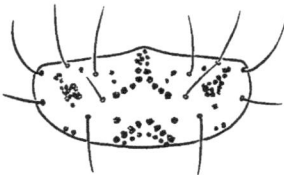

104

16(15'). Sclerotized ring around seta SD1 on abdominal segment 1 (fig. 105) · · · · · · · · · · · · · · · Galleriinae . . . 17

16'. No sclerotized ring around seta SD1 on abdominal segment 1 · · · · · · · · · · · · · · · · · Pyralinae . . . 19

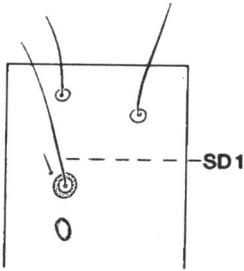

105

17(16). Prespiracular and prothoracic shields fused, setae L1 and L2 on lateral margin (fig. 106); dorsal and subdorsal setae of mesothorax on a single sclerotized plate (fig. 107) . . . . · · · · · · · · · · · · · · · · · Alpheias conspirata Heinrich

Distribution: Mexico
Hosts: pineapple

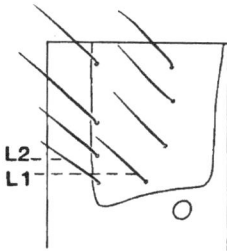

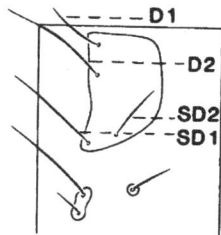

106   107

17'.   Prespiracular and prothoracic shields not fused
(fig. 108); dorsal and subdorsal setae of mesothorax on
separate pinacula (fig. 109) 18

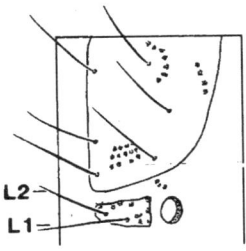
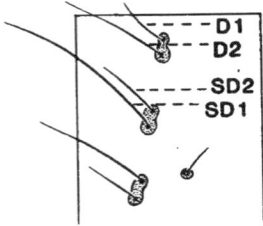

108   109

18(17'). Sclerotized rings around seta SD1 on abdominal
segments 1 and 8 not complete (fig. 110); spiracular peritremes
thicker on caudal margin (fig. 110); pinacula of setae D1 and
D2 on abdominal segments not pigmented (fig. 110) . . . . . .
. . . . Corcyra cephalonica (Stainton)

    Distribution: Cosmopolitan
    Hosts: stored vegetable products

18'.   Sclerotized rings around seta SD1 on abdominal
segments 1 and 8 complete (fig. 111); spiracular peritremes of
uniform thickness (fig. 111); pinacula of setae D1 and D2 on
abdominal segments pigmented (fig. 111) . . . .
. . . . Paralipsa gularis (Zeller)

    Distribution: Nearly cosmopolitan
    Hosts: stored vegetable products

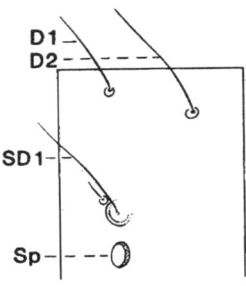
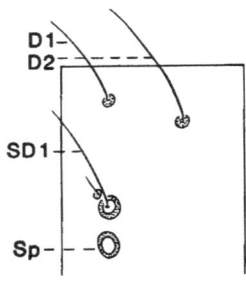

110   111

19(16'). Head with only 4 distinct ocelli present (ocelli I and II fused and ocellus VI usually missing) (fig. 112); abdominal segment 9 with one subventral seta (fig. 113) . . . . Pyralis farinalis Linnaeus

Distribution: Cosmopolitan
Hosts: dried vegetable products

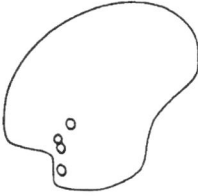

112

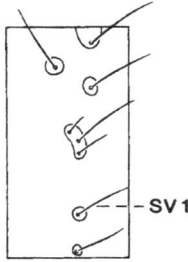

113

19'.   Head with 6 ocelli present (fig. 114); abdominal segment 9 with two subventral setae (fig. 115) . . . . Aglossa caprealis (Hübner)

Distribution: Nearly cosmopolitan
Hosts: damp grain and rotting vegetable matter (feeds on fungus)

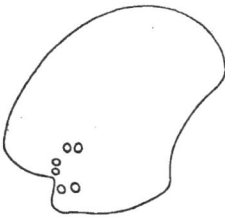

114

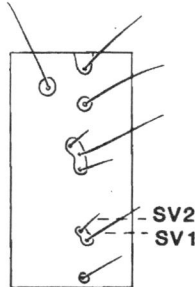

115

20(1'). A single transverse plate without setae posterior to dorsal pinacula on mesothorax (fig. 116); crochets in complete circle (fig. 117) . . . . Crambinae . . . 21

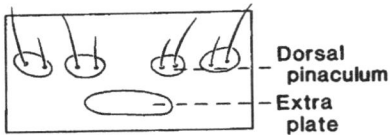

Dorsal pinaculum

Extra plate

116

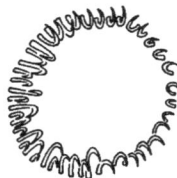

117

20'. A pair of plates without setae posterior to dorsal pinacula on mesothorax (fig. 118) or such plates absent; crochets in a mesal penellipse (fig. 119) (or may be a circle weaker on lateral edge in Lineodes integra and Udea rubigalis) (figs. 144, 146) . . . . Pyraustinae . . . 23

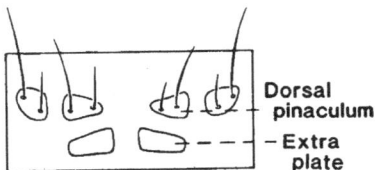

Dorsal pinaculum

Extra plate

118

119

21(20). One subventral seta on meso- and metathorax (fig. 120); body with 2 pink longitudinal stripes on each side (fig. 121); pink-pigmented area around lateral setae on proleg-bearing segments . . . . **Eoreuma loftini** (Dyar)

   Distribution: Mexico and United States
   Hosts: corn and sugarcane

21'.   Two subventral setae on meso- and metathorax (fig. 122); body with or without pigmented stripes; no pigmented area around lateral setae on proleg-bearing segments . . . 22

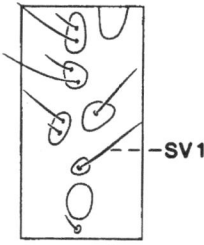

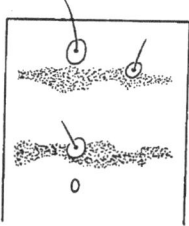

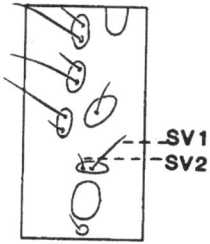

120

121

122

22(21'). Body with pinkish middorsal stripe and two lateral stripes on each side (fig. 123); setal pinacula concolorous with body . . . . **Chilo suppressalis** (Walker)

   Distribution: Japan, China, Southeast Asia to India
   Host: rice straw

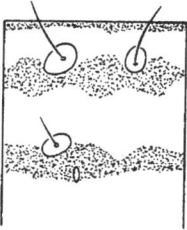

123

22'.   Body without pinkish middorsal stripe; setal pinacula
concolorous with body (winter form) or darkly pigmented (summer
form) . . . . **Diatraea** spp.

   Distribution: North, Central, and South America, and
   West Indies
   Hosts: corn, sugarcane, and sorghum

23(20'). Meso- and metathorax with a pair of nonsetal bearing
plates posterior to dorsal pinacula, also small pinacula
anterior to dorsal and subdorsal pinacula bearing microscopic
setae (fig. 124) 24

23'.   Meso- and metathorax without nonsetal bearing plates
posterior to dorsal pinacula, no small pinacula anterior to
dorsal and subdorsal pinacula (fig. 125) 25

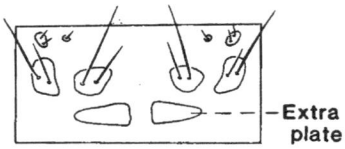

124

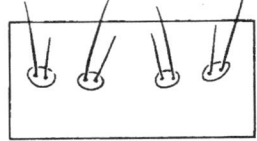

125

24(23).   Prespiracular shield of prothorax extending below and
beyond spiracle (fig. 126); an extra nonsetal bearing plate
below seta L3 on meso- and metathorax (fig. 127) and behind L1
and L2 on abdominal segments 1 to 7 (fig. 128) . . . .
. . . . **Dichocrocis** **punctiferalis** (Guenée)

   Distribution: Japan, Korea, Taiwan, and India
   Hosts: pine, chestnut, peach, and others

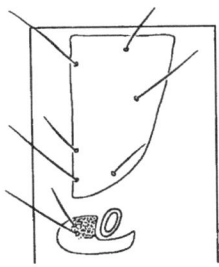

126

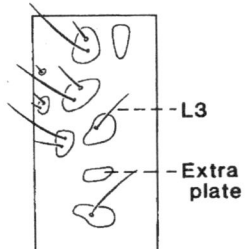

127

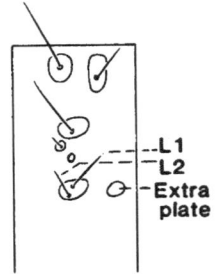

128

espiracular shield of prothorax crescent shaped
low spiracle (fig. 129); no extra nonsetal bearing
seta L3 on meso- and metathorax (fig. 130) and
d L2 on abdominal segments 1 to 7 (fig. 131) . . .
. . . . Maruca testulalis (Geyer)

ion: Africa, Asia, Australia, Mexico, Central and
erica, and Hawaii
ans, pigeon peas, and other legumes

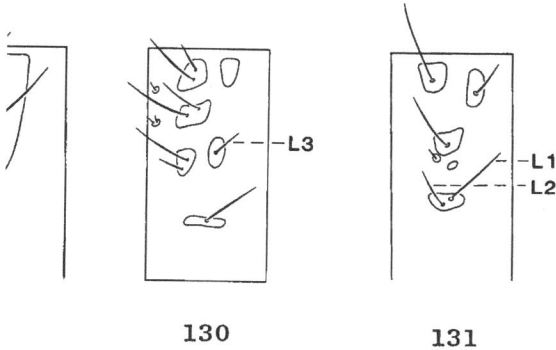

130   131

ad capsule with a shieldlike extension over base of
. 132) . . . . Ostrinia nubilalis (Hübner)

ion: Europe and United States
rn, beans, peas, and many others

ad capsule without a shieldlike extension over base
fig. 133) 26

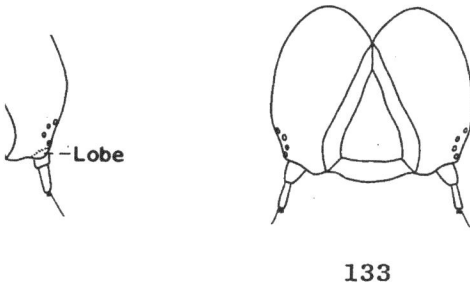

133

26(25'). Body with pinkish longitudinal stripes (fig. 134)
. . . . 27

26'.   Body without pinkish longitudinal stripes 28

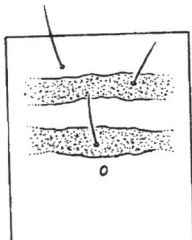

134

27(26).   Head blackish or brownish with whitish areas along
adfrontal sutures extending to vertex, seta O3 anterior to a
line joining setae L1 and O2 (fig. 135) . . . .
. . . . **_Hellula_ _rogatalis_** (Hulst)

   Distribution: Europe, North Africa, Asia, Pacific Islands,
   Mexico, West Indies, and United States
   Hosts: cabbage, mustard, radish, and turnip

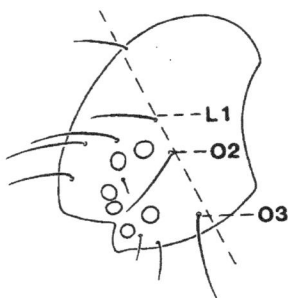

135

'.   Head pale, mottled, area along adfrontal sutures pale
t not white, seta 03 posterior to a line joining setae L1 and
(fig. 136) . . . . Hellula phidilealis (Walker)

Distribution: Mexico, Central and South America,
 West Indies, and United States
Hosts: cabbage, cauliflower, mustard, and other crucifers

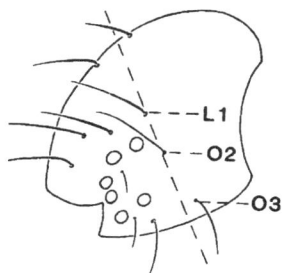

136

'26').   Prothorax with sclerotization extending from
iterolateral margin of prothoracic shield behind and below
.racle to prespiracular shield (fig. 137) . . . .
. . . . Achyra rantalis (Guenée)

Distribution: Mexico, West Indies, and United States
Hosts: alfalfa, beets, cotton, soybeans, and many others

 .   Prothorax without sclerotization extending from
terolateral margin of prothoracic shield to prespiracular
eld 29

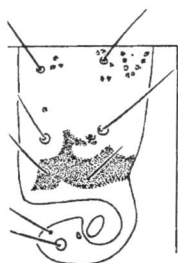

137

29(28'). Prothoracic shield broadly shaded laterally (figs. 139, 141); head yellow with dark pattern (fig. 138) 30

29'.   Prothoracic shield without dark shading laterally, head not patterned 31

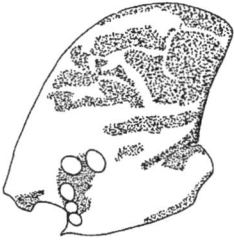

138

30(29).   Prothoracic shield with dark lateral shading extending to seta D2 (fig. 139); prespiracular shield enclosing the spiracle (fig. 139); dorsal and subdorsal pinacula of mesothorax fused (fig. 140) . . . .
. . . . Herpetogramma bipunctalis (Fabricius)

   Distribution: West Indies and United States
   Hosts: alfalfa, beets, cotton, soybeans, and many others

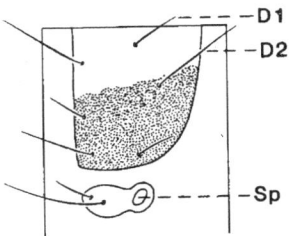

139

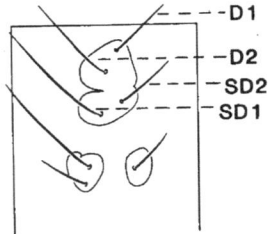

140

Prothoracic shield with dark lateral shading
lng to seta D1 (fig. 141); prespiracular shield not
lng the spiracle (fig. 141); dorsal and subdorsal
la of mesothorax not fused (fig. 142) . . . .
. . . . Pilemia periusalis (Walker)

tribution: West Indies and United States
ts: eggplant, potatoes, and tomatoes

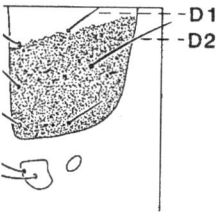

141

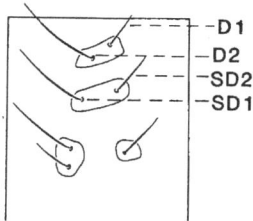

142

). Prothoracic shield with a dark reniform spot
lor to seta XD2 (figs. 143, 145) 32

Prothoracic shield without dark reniform spot
lor to seta XD2 33

. Prespiracular shield ovate (fig. 143); crochets
inal on mesal aspect (fig. 144) . . . .
. . . . Udea rubigalis (Guenée)

tribution: Mexico, West Indies, United States, and
nada
ts: celery, lettuce, spinach, and others

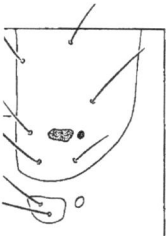

143

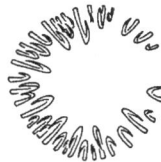

144

32'.   Prespiracular shield crescent shaped extending below
spiracle (fig. 145); crochets biordinal on mesal aspect
(fig. 146) . . . . <u>Lineodes integra</u> (Zeller)

   Distribution: Mexico and West Indies
   Hosts: eggplant and tomatoes

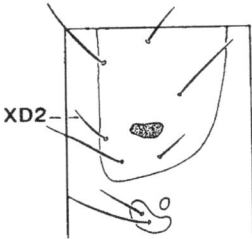

XD2

146

145

33(31'). Abdominal segment 1 with three subventral setae
(fig. 147); prothorax with seta XD2 equidistant from setae
SD1 and XD1 (fig. 148); crochets biordinal (fig. 149) . . . . .
. . . . <u>Hendecasis duplifascialis</u> Hampson

   Distribution: Philippines and Southeast Asia
   Host: jasmine

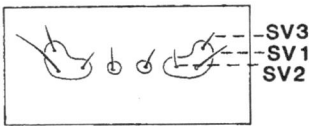

—SV3
—SV1
SV2

XD1—

XD2—

SD1—

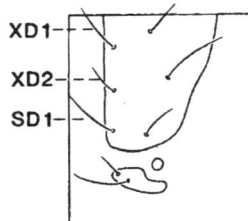

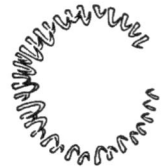

147

148

149

33'.   Abdominal segment 1 with less than three subventral setae (figs. 152, 154); prothorax with seta XD2 closer to seta SD1 than to seta XD1 (fig. 150); crochets triordinal (fig. 151)
. . . . 34

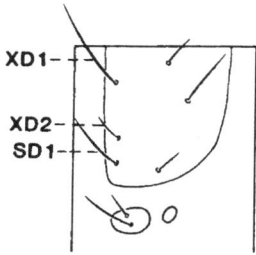

150

151

34(33'). Abdominal segment 1 with two subventral setae (fig. 152); prespiracular shield oblong (fig. 153) 35

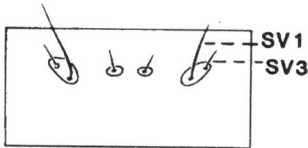

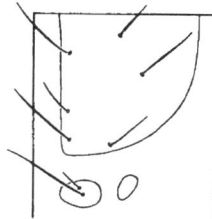

152

153

34'.   Abdominal segment 1 with one subventral seta (fig. 154); prespiracular shield crescent shaped, may extend under spiracle (fig. 155) 36

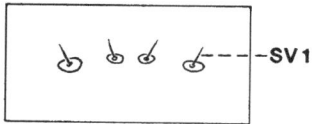

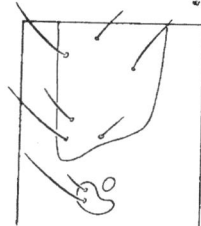

154

155

Keys for Lepidopterous Larvae - 39

35(34).   Head with a pigmented spot at genal angle (fig. 156);
mandible without a projection on lateral margin (fig. 157);
pinacula dark on early instars, pale in later instars . . . .
. . . . Diaphania nitidalis (Stoll)

   Distribution: Mexico, Central and South America,
   West Indies, United States, and Canada
   Hosts: squash, cantaloupe, cucumbers, and gourds

35'.   Head without pigmented spot at genal angle; mandible
with a projection on lateral margin (fig. 158); pinacula
concolorous with body in all instars . . . .
. . . . Diaphania hyalinata (Linnaeus)

   Distribution: Mexico, Central America, Northern South
   America, West Indies, and Eastern United States
   Hosts: squash, cucumbers, cantaloupe, gourds, and pumpkins

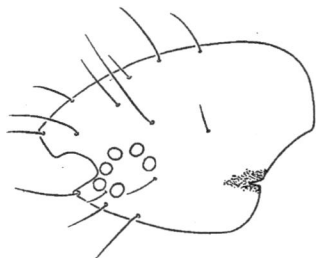

156   158
                    157

36(34'). Head, prothoracic shield, and body pinacula brownish
yellow; pinaculum of seta D1 on abdominal segments 2 to 8 with
dark spot on anterior margin (fig. 159) . . . .
. . . . Leucinodes orbonalis (Guenée)

   Distribution: Africa and Southeast Asia
   Hosts: eggplant, tomatoes, potatoes, and other solanaceous
   plants

159

36'.   Head and prothoracic shield pale yellow, pinacula concolorous with body; pinaculum of seta D1 on abdominal segments 2 to 8 without spots . . . .
· · · · · · · · · · · · · · · Neoleucinodes elegantalis (Guenée)

   Distribution: Mexico, Central and South America, and
   West Indies
   Hosts: eggplant, tomatoes, and other solanaceous plants

1.   Seta D1 of abdominal segment 9 cephalad of and equidistant from setae D2 and SD1, setae D1 and SD1 on separate pinacula (fig. 160) . . . . Tortricinae . . . 2

1'.   Seta D1 of abdominal segment 9 closely associated with seta SD1 on a single pinaculum (fig. 161),
· Olethreutinae . . . 3

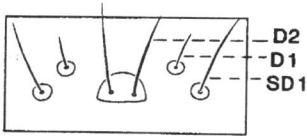   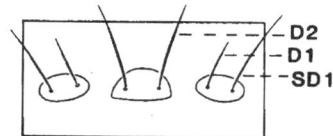

          160   161

2(1).   Head, prothoracic shield, and prothoracic pinacula yellowish, shield may be edged laterally and posteriorly with brown . . . . Platynota stultana (Walsingham)

   Distribution: Mexico and United States
   Hosts: tomatoes, peppers, and many others

2'.   Head, prothoracic shield, and prothoracic pinacula brown or blackish, shield and head may show faint pattern . .
· · · · · · · · · · · · · · · · · Platynota rostrana (Walker)

   Distribution: Mexico, Central America, West Indies, and
   United States
   Hosts: cotton, banana, and many others

(1').   Anal fork present (fig. 162); three lateral setae
f abdominal segment 9 on one pinaculum (fig. 163) · · · · · · 4

'.   Anal fork absent; seta L3 of abdominal segment 9
sually on its own pinaculum separate from that of setae LI
nd L2 (fig. 164) · 7

162

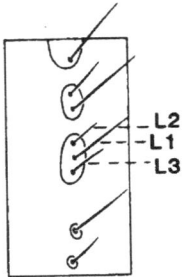

163

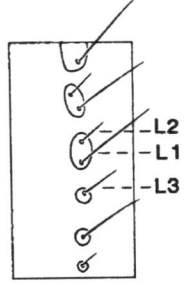

164

(3).   Prespiracular shield of prothorax elongated extending
elow and beyond spiracle (fig. 165) · · · · · · · · · · · · ·
· · · · · · · · · · · · · Cryptophlebia leucotreta (Meyrick)

   Distribution: Central and South Africa
   Hosts: okra, orange, Capsicum sp., and many others

'.   Prespiracular shield of prothorax not extending
elow spiracle (fig. 166) · · · · · · · · · · · · · · · · · · 5

165

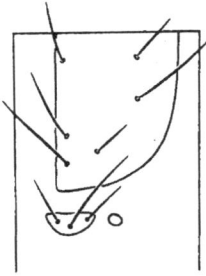

166

5(4').   Body pinacula large and brown (fig. 167); prothoracic
shield (fig. 168) and anal shield (fig. 169) patterned as
illustrated . . . . **Pammene fasciana** (Linnaeus)

    Distribution: Europe
    Hosts: acorns and chestnuts

5'.   Body pinacula smaller and pale; prothoracic and anal
shields not patterned as above 6

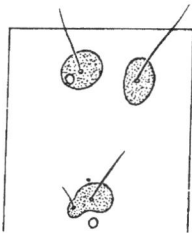

167

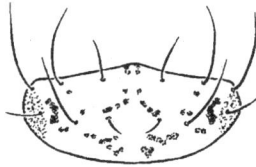

168

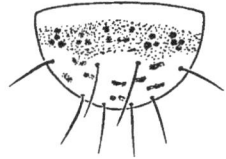

169

6(5').   Head with black band extending from postgenal
juncture to seta O2 (fig. 170); prothorax with seta SD2
dorsocaudad of SD1, lateral setae almost in horizontal line
(fig. 171) . . . . **Epinotia aporema** (Walsingham)

    Distribution: Mexico, Central and South America,
    West Indies, and United States
    Hosts: string beans and other legumes

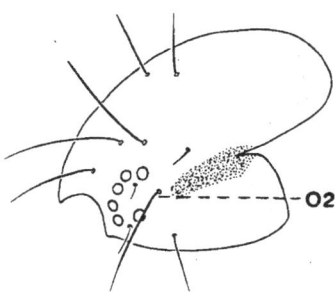

170

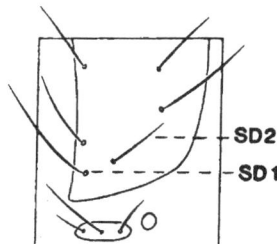

171

6'.   Head with some dark color at postgenal juncture but
not extending to seta O2 (fig. 172); prothorax with seta SD2
directly caudad of SD1, lateral setae in more triangular
arrangement (fig. 173) . . . . *Grapholita* spp.

   Distribution: Europe, East Asia, Australia, Mexico,
     South America, and United States
   Hosts: apple and other pomes, plums and other drupes, and
     berries

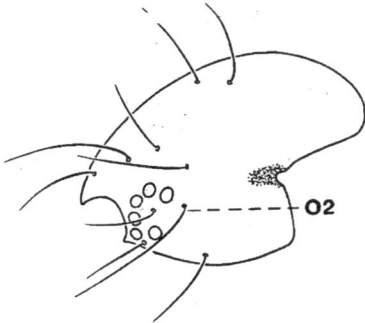

172

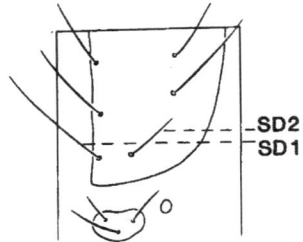

173

7(3').   Head yellowish brown, usually distinctly patterned
with dark color (fig. 174), prothoracic shield (fig. 175) and
anal shield (fig. 176) yellowish brown with dark patterns as
illustrated . . . . *Cydia pomonella* (Linnaeus)

   Distribution: Nearly cosmopolitan
   Hosts: apples, pears, quince, and walnut

7'.   Head yellow, possibly faintly patterned; prothoracic
and anal shields yellow 8

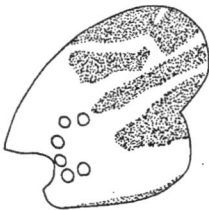

174

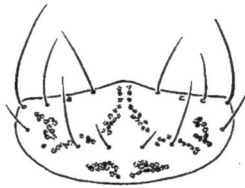

175

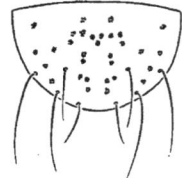

176

8(7').   Head clear yellow; body always white . . . .
. . . . Cydia splendana (Hübner)

    Distribution: Southern Europe
    Host: chestnuts

8'.   Head yellow with some indication of faint pattern or
some dark color at postgenal juncture; body with pink pigment
in fresh specimens . . . . Cydia spp.

    Distribution: Europe and East Asia
    Host: chestnuts

This family never has more than two lateral setae on abdominal
segment 9. The ventral setae are always farther apart on the
9th abdominal segment than on the 8th . . . .
. . . . Lorita abornana chatka Busck

    Distribution: Mexico
    Host: peppers

Many species in this family have rings around seta SD1 on
abdominal segments 1 to 7. The subfamily Blastobasinae always
has the submental pit (fig. 29) and always has three subventral
setae on abdominal segment 1 . . . . . Species of Blastobasidae

    Distribution: Worldwide
    Hosts: garlic, coffee, acorns, and many others

1.   Proleg-bearing segments of abdomen (3-6) usually with
three (number variable) extra unnamed setae between the dorsal
and subdorsal setae (fig. 177); crochets in two uniordinal
crossbands (fig. 178) . . . . _Dyspessa_ _ulula_ (Borkhausen)

Distribution: Europe
Host: garlic

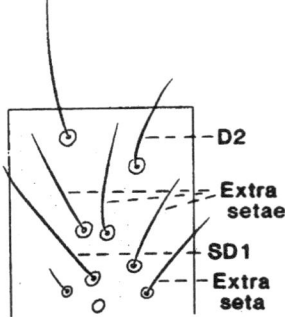

177

178

1'.   Proleg-bearing segments of abdomen without three
extra setae between the dorsal and subdorsal setae (fig.
179); crochets in a biordinal laterally elongated circle
(fig. 180) . . . . _Cossus_ _cossus_ (Linnaeus)

Distribution: Europe
Hosts: wood products

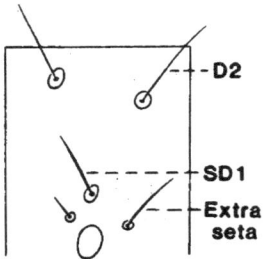

179

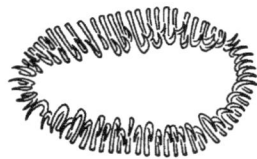

180

The character of setae SD2 being almost in line with setae XD2 and SD1 appears to be consistent for this family · · · · · · ·
· · · · · · · · · · · · · · · · **Argyresthia conjugella** (Zeller)

    Distribution: Europe
    Hosts: apples and sorbus berries

Setae L1 and L2 on one pinaculum, and seta L3 separate on abdominal segment 9; three subventral setae on abdominal segment 1 · · · · · · · · · · · · · · · · · · **Pyroderces** spp·

    Distribution: Mexico, West Indies, and United States
    Hosts: corn, cotton, and many rotting and dried fruits

1.   Abdominal segment 8 with seta L3 above the level of setae L1 and L2, spiracle toward back of segment (fig. 181); prothorax with large prespiracular shield extending below the spiracle (fig. 182) · · · · · · **Stenoma catenifer** Walsingham

    Distribution: Mexico, Central and South America
    Host: avocado

1'.   Abdominal segment 8 with seta L3 below the level of setae L1 and L2, spiracle in usual position (fig. 183); prothorax without prespiracular shield extending below the spiracle · 2

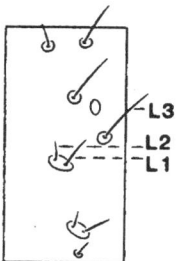
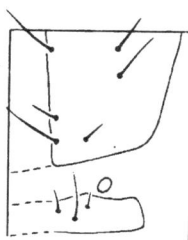
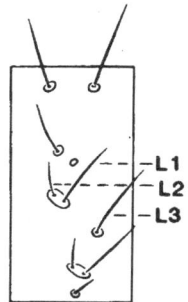

181   182   183

2(1').   Head with 2 ocelli present (fig. 184); submentum with large oval pit (fig. 185) . . . <u>Endrosis</u> <u>sarcitrella</u> (Linnaeus)

   Distribution: Nearly cosmopolitan
   Hosts: bulbs and decaying fruits

2'.   Head with 4 ocelli apparent (ocelli I and II fused and ocelli III and IV fused) (fig. 186); submentum without large oval pit . . . <u>Hofmannophila</u> <u>pseudospretella</u> (Stainton)

   Distribution: Nearly cosmopolitan
   Hosts: bulbs, stored vegetable products, and many others

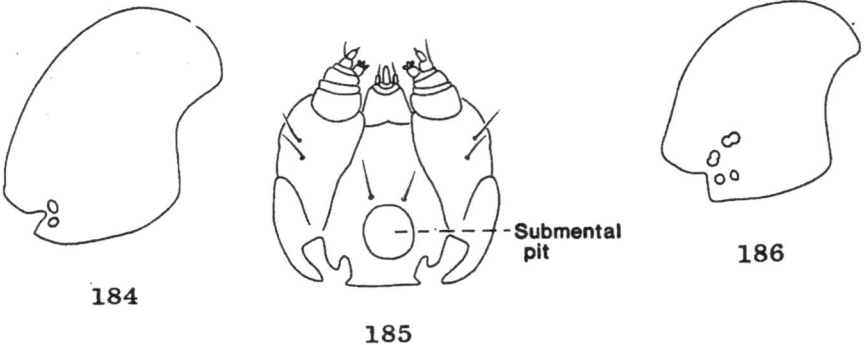

184

185

-Submental pit

186

1.   Abdominal prolegs rudimentary, with only 2 to 4 crochets (fig. 187) . . . . . . <u>Sitotroga</u> <u>cerealella</u> (Olivier)

   Distribution: Nearly cosmopolitan
   Hosts: stored grain

1'.   Abdominal prolegs normal, each proleg with more than 4 crochets 2

187

2(1').   Prothorax with prespiracular shield enclosing the
spiracle, lateral setae in a linear arrangement (fig. 188);
crochets of anal prolegs interrupted at center (fig. 189); anal
fork present (fig. 189) . . . . . . <u>Anarsia lineatella</u> Zeller

    Distribution: Europe, Mexico, Central and South America,
     and United States
    Hosts: peach, pear, almond, cherry, and other stone fruits

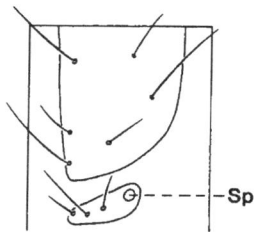

188

189

2'.   Prothorax with prespiracular shield not enclosing the
spiracle, lateral setae in a triangular arrangement (fig. 190);
crochets of anal prolegs not interrupted at center (fig. 191);
anal fork absent (fig. 191) 3

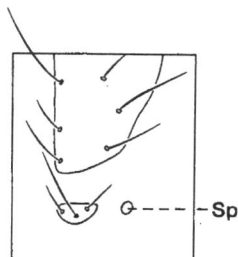

190

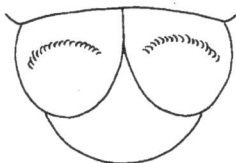

191

Head with adfrontal setae widely separated, seta
level of apex of front (fig. 192); abdominal prolegs
ochets in a uniordinal penellipse (fig. 193) . . . . .
. . . . Pectinophora gossypiella (Saunders)

ribution: India, Egypt, Mexico, Central and
th America, West Indies, and United States
s: cotton, okra, and other malvaceous plants

Head with adfrontal setae close together, seta Adf2
low apex of front (fig. 194); abdominal crochets in a
e circle 4

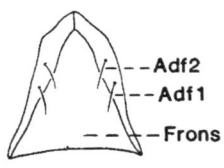

--Adf2

--Adf1

--Frons

--Adf2

--Adf1

--Frons

193   194

Prothoracic shield pale, with dark shading along
or margin (fig. 195) . . . .
. . . . Keiferia lycopersicella (Walsingham)

ribution: Mexico, Central and South America,
t Indies, Hawaii, and United States
s: tomato, eggplant, potato, and other solanaceous
its

Prothoracic shield uniformly brown or blackish . . 5

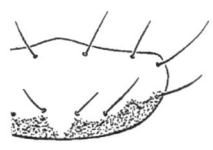

195

5(4'). Line joining setae L1 and O2 tangent to or passing through ocellus I (fig. 196); lateral setae of abdominal segment 9 in a nearly vertical line (fig. 197); legs pale . . . . **Symmetrischema** capsicum (Bradley & Povolny)

Distribution: Mexico, Central and South America, West Indies, and United States
Hosts: pepper and tomato

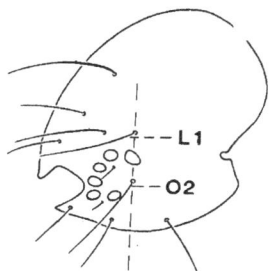

196

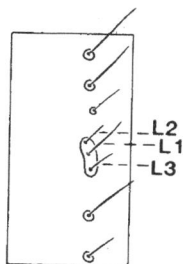

197

5'. Line joining setae L1 and O2 posterior to ocellus I (fig. 198); lateral setae of abdominal segment 9 in a triangular arrangement (fig. 199); legs dark . . . . **Phthorimaea** operculella (Zeller)

Distribution: Nearly cosmopolitan
Hosts: potatoes, tomatoes, stored tobacco, and other solanaceous plants

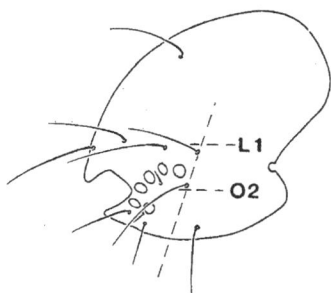

198

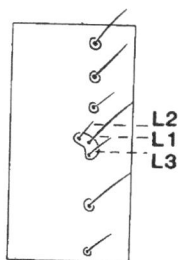

199

The pupae in this family are always enclosed in loose net; Europe and Hawaii; leeks . . . . <u>Acrolepia</u> <u>assectella</u> (Zeller)

1.   Anal prolegs longer than broad, few crochets (fig. 200); abdominal segment 9 with dorsal, subdorsal, and lateral setae all widely separated, seta SD1 distinctly thin and hairlike (fig. 201) . . . . . . <u>Plutella</u> <u>xylostella</u> (Linnaeus)

   Distribution: Nearly cosmopolitan
   Hosts: cabbage and other crucifers

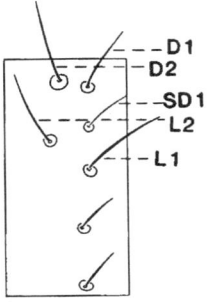

200

201

1'.   Anal prolegs short, many crochets (fig. 202); abdominal segment 9 with dorsal and subdorsal setae on one continuous pinaculum and the lateral setae on another, seta SD1 not hairlike (fig. 203) . . . . <u>Prays</u> spp.

   Distribution: Europe, East Asia, and Hawaii
   Hosts: citrus, olives, and pelea berries

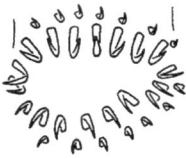

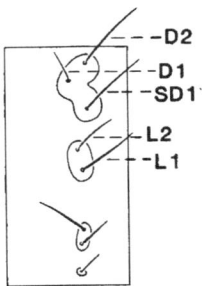

202

203

Abdominal prolegs with rows of minute recurved
anterior and posterior to the crochets (fig. 204) or
or aspect only (fig. 205); abdominal segment 1 with
ventral setae (fig. 206) 2

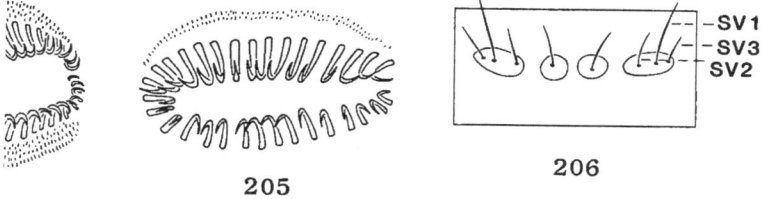

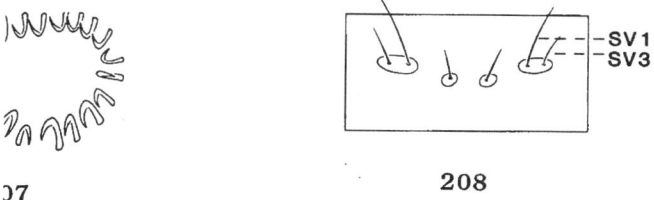

205

206

1

Abdominal prolegs without minute spinules adjacent to
ets (fig. 207); abdominal segment 1 with two sub-
etae (fig. 208) 4

07

208

Abdominal prolegs with rows of spinules anterior to
ets (fig. 209); head with only one ocellus (fig. 210)
. . . . Setomorpha rutella Zeller

bution: Nearly cosmopolitan
dried tobacco, cottonseed, and many stored plant
cts

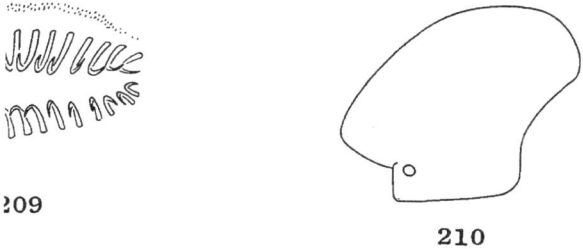

209

210

Keys for Lepidopterous Larvae - 53

lominal prolegs with rows of spinules both anterior
: to the crochets (figs. 213, 216); head with
: six ocelli (figs. 211, 214) 3

ıd with 6 ocelli (fig. 211); prespiracular shield
.racle and is fused to prothoracic shield (fig.
ıal prolegs with space between the spinules and
;. 213) . . . . _Acrolophus_ spp.

.on: Mexico, Central and South America
ımeliads, orchids, and others

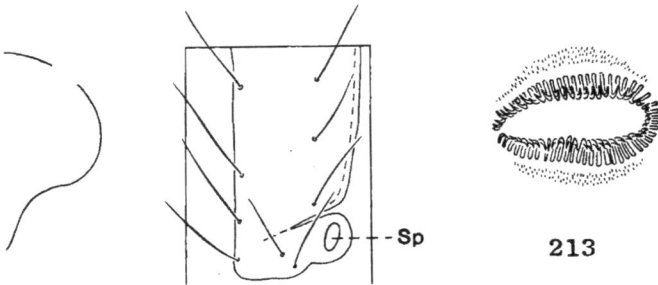

-Sp

212   213

d with 2 ocelli (fig. 214); prespiracular shield
racle but is not fused to prothoracic shield
bdominal prolegs without space between the
crochets (fig. 216) . . . _Opogona_ _sacchari_ (Bojer)

on: Southern Europe, Africa, Brazil, and
es
arcane, banana, bulbs, and many others

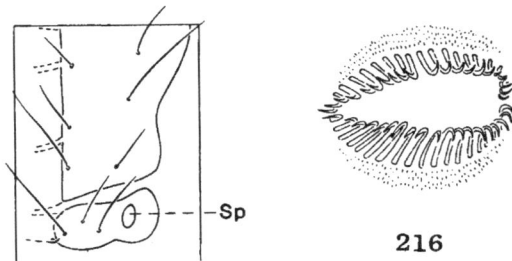

-Sp

215   216

4(1').   Meso- and metathorax with two subventral setae
(fig. 217); abdominal segment 9 with two lateral setae and one
subventral seta (fig. 218); head with no ocelli present . . .
. . . . Tineola bisselliella (Hummel)

    Distribution: Cosmopolitan
    Hosts: wool, hair, feathers, and other animal products

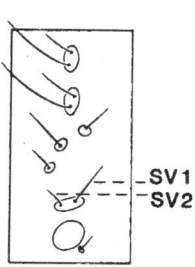 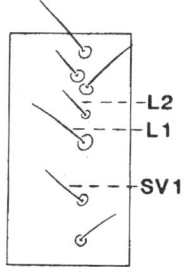

        217   218

4'.   Meso- and metathorax with one subventral seta
(fig. 219); abdominal segment 9 with three lateral setae and
two subventral setae (fig. 220); head with 5 or 6 ocelli
present 5

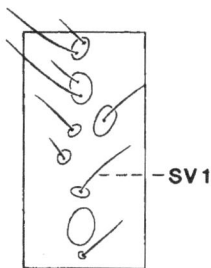 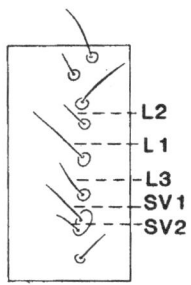

        219   220

5(4').   Head with 5 ocelli (fig. 221); prespiracular shield
extends behind spiracle (fig. 222) . . . .
. . . . Lepidobregma minuscula (Walsingham)

   Distribution: West Indies, Hawaii, and United States
   Hosts: pineapple, banana, and many others (scavenger)

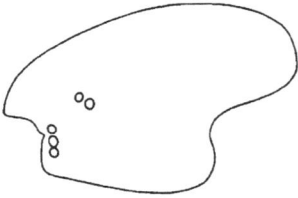

221

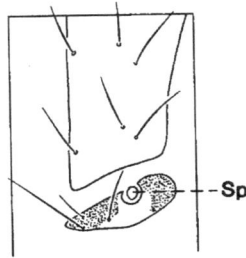

222

5'.   Head with 6 ocelli (fig. 223); prespiracular shield
does not extend behind spiracle . . . .
. . . . Nemapogon granella (Linnaeus)

   Distribution: Nearly cosmopolitan
   Hosts: mushrooms, stored grain, and dried fruits

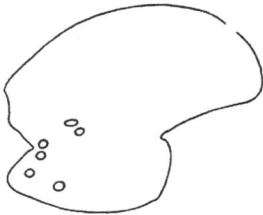

223

Aitken, A. D. A key to the larvae of some species of Phycitinae (Lepidoptera, Pyralidae) associated with stored products, and of some related species. Bull. Entomol. Res. 54(2):175-188; 1963.

Allyson, S. North American larvae of the genus Loxostege Hübner (Lepidoptera: Pyralidae: Pyraustinae). Can. Entomol. 108(1): 89-104; 1976.

_____. Last instar larvae of Pyraustini of America north of Mexico (Lepidoptera: Pyralidae). Can. Entomol. 113(6):463-518; 1981.

Baker, C. R. B. Notes on the larvae and pupae of two fruit moths, Grapholita funebrana Treitschke and G. molesta Busck (Lepidoptera: Olethreutidae). Proc. R. Entomol. Soc. London Ser. A, 38:212-222; 1963.

Beck, H. Die Larvalsystematik der Eulen (Noctuidae). Abh. Larvalsys. Insekten 4; 1960.

Bollmann, H. G. Die Raupen mitteleuropäischer Pyraustinae (Lepidoptera: Pyralidae). Beitr. Entomol. 5:521-639; 1955.

Busck, A. The pink bollworm, Pectinophora gossypiella. J. Agric. Res. 9(10):333-370; 1917.

Capps, H. W. Description of the larva of Keiferia peniculo Heinrich, with a key to the larvae of related species attacking eggplant, pepper, potato and tomato in the United States (Lepidoptera: Gelechiidae). Ann. Entomol. Soc. Am. 39:561-563; 1946.

_____. An illustrated key for identification of larvae of the cotton-pest species of Pectinophora Busck and Platyedra Meyrick (Lepidoptera, Gelechiidae). Bull. Entomol. Res. 49(4): 631-632; 1958.

_____. Keys for the identification of some lepidopterous larvae frequently intercepted at quarantine. ARS 30-20-1. Washington, DC: Agricultural Research Service, U.S. Department of Agriculture; 1963.

Chu, H. F.; Fang, C. L.; Wang, L. Y. Fauna of Chinese economic insects. Fasc. 7. Noctuidae (immature stages). Peking: Science Press; 1963 (In Chinese).

Crumb, S. E. The nearctic budworms of the lepidopterous genus Heliothis. U.S. Natl. Mus. Proc. 68(No. 2617, art. 16); 1926.

_____. The armyworms. Brooklyn Entomol. Soc. Bull. 22:41-45; 1927.

_____. Tobacco cutworms. U.S. Dep. Agric. Tech. Bull. 88; 1929.

_____. The more important climbing cutworms. Brooklyn Entomol. Soc. Bull. 27:73-100; 1932.

_____. The larvae of the Phalaenidae. U.S. Dep. Agric. Tech. Bull. 1135; 1956.

Davis, E. G.; Horton, J. R.; Gable, C. H.; Blanchard, R. A.; and Heinrich, C. The southwestern corn borer. U.S. Dep. Agric. Tech. Bull. 388; 1933.

Eichlin, T. D. Guide to the adult and larval Plusiinae of California (Lepidoptera: Noctuidae). Occas. Pap. Entomol., State Calif. Dep. Food Agric. No. 21; 1975.

Eichlin, T. D.; Cunningham, H. B. The Plusiinae (Lepidoptera: Noctuidae) of America north of Mexico emphasizing genitalic and larval morphology. U.S. Dep. Agric. Tech. Bull. 1567; 1978.

Forbes, W. T. M. The Lepidoptera of New York and neighboring states. Cornell Univ. Agric. Exp. Stn. Mem. 68; 1923.

Fracker, S. B. The classification of lepidopterous larvae. Ill. Biol. Monogr. 2(1); 1915.

Godfrey, G. L. A review and reclassification of larvae of the subfamily Hadeninae (Lepidoptera, Noctuidae) of America north of Mexico. U.S. Dep. Agric. Tech. Bull. 1450; 1972.

Hasenfuss, I. Die Larvalsystematik der Zunsler (Pyralidae). Abh. Larvalsys. Insekten 5; 1960.

Heinrich, C. Note on the European corn borer (Pyrausta nubilalis Hübner) and its nearest American allies, with description of larvae, pupae, and one new species. J. Agric. Res. 18(3):171-178; 1919.

_____. Some Lepidoptera likely to be confused with the pink bollworm. J. Agric. Res. 20(11):807-836; 1921.

E. The larvae of the Lepidoptera associated with
:oducts. Bull. Entomol. Res. 34:163-212; 1943.

ι the homology and nomenclature of the setae of
:rous larvae with some notes on the phylogeny of the
:ra. Trans. R. Entomol. Soc. London 97:1-37; 1946.

: larvae of the species of Tineidae of economic
:e. Bull. Entomol. Res. 47(2):251-346; 1956.

:. E.; Haley, W. E.; Loftin, U. C.; Heinrich, C.
:cane moth borer in the United States. U.S. Dep.
:ch. Bull. 41; 1928.

Early stages of Japanese moths in colour. Rev. ed.
ιpan: Hoikusha Publishing Co., Ltd. Vol. 1; 1973.
.975.

[abeck, D. H. Descriptions of the larvae of
:a _sunia_ and S. _latifascia_ with a key to the mature
:a larvae of the eastern United States (Lepidoptera:
:). Ann. Entomol. Soc. Am. 69(4):585-588; 1976.

R. Larvae of the North American Olethreutidae
.era). Can. Entomol. Suppl. 10; 1959.

·vae of the North American Tortricinae (Lepidoptera:
ae). Can. Entomol. Suppl. 28; 1962.

.l sketches of some Microlepidoptera, chiefly North
Mem. Entomol. Soc. Can. 88:1-83; 1972.

aya, O. I. The larvae of Noctuidae, their biology
ology (with keys). Minsk, Byelorussia, USSR: Academy
e, Byelorussia (Department of Zoology and
ogy), Science and Technology; 1967 (In Russian).

A classification of the Lepidoptera based on
s of the pupa. Bull. Ill. State Lab. Nat. Hist.
159; 1916.

H. Systematics of immature phycitines (Lepidoptera:
) associated with leguminous plants in the southern
ates. U.S. Dep. Agric. Tech. Bull. 1589; 1979.

T. Identifications of lepidopterous larvae
cotton. Calif. Dep. Agric., Bur. Entomol. Spec.
; 1961.

Peterson, A. Larvae of insects. Part 1. Lepidoptera and plant infesting Hymenoptera. Columbus, OH: Alvah Peterson; 1956.

Swatschek, B. Die Larvalsystematik der Wickler (Tortricidae und Carposinidae). Abh. Larvalsys. Insekten 3; 1958.

Werner, K. Die Larvalsystematik einiger Kleinschmetterlingfamilien (Hyponomeutidae, Orthoteliidae, Acrolepiidae, Tineidae, Incurvariidae und Adelidae). Abh. Larvalsys. Insekten 2; 1958.

Zimmerman, E. C. Insects of Hawaii. Vol. 9. Microlepidoptera. Parts 1 and 2. Honolulu: The Univ. Press of Hawaii; 1978.

abornana chatka, Lorita 45
Achyra 35
Acrolepia 52
Acrolepiidae 9, 52
Acrolophus 54
Aglossa 29
Alpheias 27
Amyelois 22, 23
Anagasta 25
Anarsia 49
Ancylostomia 20
aporema, Epinotia 43
Argyresthia 47
Argyresthiidae 7, 47
Ascia 11
assectella, Acrolepia 52
Autographa 13

bipunctalis, Herpetogramma 36
bisinuella, Moodna 21
bisselliella, Tineola 55
Blastobasidae 6, 45
boeticus, Lampides 12
brassicae, Mamestra 16
brassicae, Pieris 11

Cadra 25, 26
calidella, Cadra 26
caprealis, Aglossa 29
capsicum, Symmetrischema 51
Carposina 17
Carposinidae 17
catenifer, Stenoma 47
cautella, Cadra 25
cephalonica, Corcyra 28
ceratoniae, Ectomyelois 23
cerealella, Sitotroga 48
Chilo 31
Cochylidae 6, 45
conjugella, Argyresthia 47
conspirata, Alpheias 27
Copitarsia 17
Corcyra 28
Cosmopterigidae 8, 47
Cossidae 7, 46
Cossus 46
cossus, Cossus 46
Cryptophlebia 42
Cydia 44, 45

Diaphania . . . .
Diatraea . . . .
Dichocrocis . . . .
duplifascialis, Hendecasis . .
Dyspessa . . . .

Ectomyelois . . . .
Elasmopalpus . . . .
elegantalis, Neoleucinodes . .
elutella, Ephestia . . . . . .
Eoreuma . . . .
Endrosis . . . .
Ephestia . . . .
Epinotia . . . .
Etiella . . . .
exigua, Spodoptera . . . . . .

farinalis, Pyralis . . . . . .
fasciana, Pammene . . . . . .
figulilella, Cadra . . . . . .
frugiperda, Spodoptera . . . .
Fundella . . . .

gamma, Autographa . . . . . .
Gelechiidae . . . .
Geometridae . . . .
gossypiella, Pectinophora . .
grandella, Hypsipyla . . . . .
granella, Nemapogon . . . . .
Grapholita . . . .
gularis, Paralipsa . . . . . .

Helicoverpa . . . .
Heliothis . . . .
Hellula . . . .
Hendecasis . . . .
Herpetogramma . . . .
Hesperiidae . . . .
Hofmannophila . . . .
hyalinata, Diaphania . . . . .
Hypsipyla . . . .

Idaea . . . .
integra, Lineodes . . . . . .
interpunctella, Plodia . . . .

Keiferia . . . .
kuehniella, Anagasta . . . . .

Lampides 12
Lepidobregma 56
Leucinodes 40
leucotreta, Cryptophlebia 42
lignosellus, Elasmopalpus 19
lineatella, Anarsia 49
Lineodes 30, 38
loftini, Eoreuma 31
Lorita 45
Lycaenidae 2, 12
lycopersicella, Keiferia 50

Mamestra 16
Maruca 33
melinus, Strymon 12
minuscula, Lepidobregma 56
monuste, Ascia 11
Moodna 21

napi, Pieris 10
Nemapogon 56
Neoleucinodes 41
niponensis niponensis, Carposina 17
nitidalis, Diaphania 40
ni, Trichoplusia 13
Noctuidae 13
nubilalis, Ostrinia 33

Oecophoridae 8, 47
operculella, Phthorimaea 51
Opogona 54
orbonalis, Leucinodes 40
Ostrinia 33

Fammene 43
Paralipsa 28, 50
Pectinophora 50
pellucens, Fundella 21
Peridroma 16
periusalis, Pilemia 37
phidilealis, Hellula 35
Phthorimaea 51
Pieridae 2, 10
Pieris 10, 11
Pilemia 37
Platynota 41
Plodia 24
Plutella 52
Plutellidae 9, 52

pomonella, Cydia 44
Prays 52
pseudospretella, Hofmannophila 48
punctiferalis, Dichocrocis 32
Pyralidae 4, 18
Pyralis 29
Pyroderces 47

rantalis, Achyra 35
rapae, Pieris 10
rogatalis, Hellula 34
rostrana, Platynota 41
rubigalis, Udea 30, 37
rutella, Setomorpha 53

sacchari, Opogona 54
sarcitrella, Endrosis 48
saucia, Peridroma 16
Setomorpha 53
Sitotroga 48
splendana, Cydia 45
Spodoptera 15
Stenoma 47
stercorea, Ancylostomia 20
Strymon 12
stultana, Platynota 41
suppressalis, Chilo 31
Symmetrischema 51

testulalis, Maruca 33
Tineidae 9, 53
Tineola 55
Tortricidae 6, 41
transitella, Amyelois 22, 23
Trichoplusia 13

Udea 30, 37
ulula, Dyspessa 46

virescens, Heliothis 14

Xestia 16, 17
xylostella, Plutella 52

zea, Helicoverpa 14
zinckenella, Etiella 18, 26

☆U.S. Government Printing Office : 1986 - 490-918/40176

CPSIA information can be obtained
at www.ICGtesting.com
Printed in the USA
BVHW040840070219
539714BV00015B/135/P